Is Rubik's Cube™ driving you bonkers? Then this is what you've been waiting for.

Unlike other solutions, this solution is both easy to follow and is deliberately presented without reference to the colors on the faces of the cube. (Take a closer look, all cubes are not colored alike!)

Try it with just a few hints (pages 21-23) or with the quick and complete, step-by-step solution which follows (pages 31-53).

Amaze your friends! Master that infernal cube once and for all!

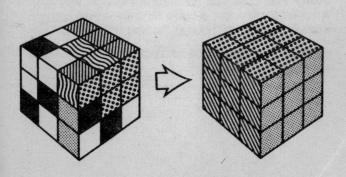

About the author
James G. Nourse

Jim is in the Department of Chemistry at Stanford University in California.

He insists that the problem of solving Rubik's Cube is very similar to the daily problems he faces in Chemistry.

Jim bought a cube as a 1980 Christmas gift for someone and became so enthralled with it that he kept it for himself. At the time, many of his colleagues were working unsuccessfully on solving the cube. He felt very certain that he was uniquely qualified to solve the enigma of Rubik's Cube because of his academic background. (Jim is a group-theory mathematician as well as holding a doctorate in Chemistry studying molecular structure and movement of particles.)

During that Christmas vacation he began writing his simplified solution to the cube. His cube is battered and worn from the many hours of twisting and twirling.

He presently uses his cube as a pacifier. He says it beats drinking or nail biting.

THE
SIMPLE SOLUTION
TO RUBIK'S CUBE™

BY

JAMES G. NOURSE

ILLUSTRATED BY DUSAN KRAJAN

BANTAM BOOKS

TORONTO / NEW YORK / LONDON / SYDNEY

To Cindy

Acknowledgments

This book would have withered away
and died had it not been for the efforts of
Paul N. Weinberg of STORC Enterprises. It
was his marketing skills that brought it
back to life and demonstrated to me its
potential. Many of his ideas substantially
improved the book and have been incor-
porated into it. The value of a skilled
partner is incalculable.

The idea for the game of displaying words
on the cube originated with Jack Looney,
to whom I am grateful.

THE SIMPLE SOLUTION TO RUBIK'S CUBE™
*A Bantam Book / Published by arrangement with
Paul N. Weinberg and STORC Enterprises,
Mountain View, CA 94043*

Bantam edition / June 1981

ISBN 0-553-14017-5

Published simultaneously in the United States and Canada

*Bantam Books are published by Bantam Books, Inc. Its trademark, con-
sisting of the words "Bantam Books" and the portrayal of a bantam, is
Registered in U.S. Patent and Trademark Office and in other countries. Mar-
ca Registrada. Bantam Books, Inc., 666 Fifth Avenue, New York, New York
10103.*

PRINTED IN THE UNITED STATES OF AMERICA

17 16 15 14

Contents

Introduction 7

About the Rubik's Cube 11

Summary of Solution 18

Notation and Terminology 24

Complete Solution 31

    Step 1 (Top Edges) 31

    Step 2 (Top Corners) 35

    Step 3 (Vertical Edges) 38

    Step 4 (Bottom Corners) 42

    Step 5 (Bottom Edges) 47

Afterword 54

Other Games to Play 56

    Speed Cubing 56

    Competition 56

    Answers 63

INTRODUCTION

If you're the owner of Rubik's Cube or a similar cube, you're probably finding that no matter how hard you try and how much time you spend fiddling with it, the cube still looks like a pile of colored confetti.

What you need is a solution to this puzzle that is not only guaranteed to work but also is simple enough to understand even if you're not a mathematical genius.

The problem of restoring a cube to solid colors on all 6 faces is very difficult–far more difficult than nearly any other popular puzzle. As a result, many available solutions are also so difficult that understanding them is no easier than solving the puzzle itself. The purpose of this book is to present a complete, quick, step-by-step solution that is both straightforward and effective. No knowledge of mathematics is assumed or necessary.

Features that distinguish this solution from others are the following:

- **It is guaranteed to work.**
- **It offers clear and steady progress.**
- **It presents few and uncomplicated decisions.**
- **It allows for error.**
- **It features easy-to-memorize sequences of moves.**
- **It is independent of color.**

7

It is guaranteed to work.

No matter how badly your cube is scrambled, this solution provides every step from beginning to end to restore it to solid colors on all 6 faces. Every step is provided in an easy-to-follow sequence. The solution does not involve aimless moves, nor is it necessary that the cube be half solved to begin making moves.

It offers clear and steady progress.

The solution starts with the top face (which you choose) and proceeds steadily to the bottom face. This solution has been designed so that very little of the previously completed portion of the cube need be disturbed to proceed. In other words, after you have completed one face, you will need to disturb very little of it to continue, and then only temporarily.

In general, no more than two previously placed cubes will be more than one move away from their proper position. It is, however, very important to the solution that you realize (and accept) the fact that temporary disruption of hard-won progress is necessary. Indeed, it is probably the most difficult aspect of solving the cube for the first time.

It presents few and uncomplicated decisions.

At some of the steps in the solution, it is necessary to make decisions in order to proceed to the correct next step. This solution has been designed so that these decisions are as few and uncomplicated as possible. When such decisions are necessary, illustrations are frequently used to facilitate understanding. You will have to match the pattern of colors that occur on your cube to one of several illustrated patterns in order to decide on the next step.

It allows for error.

The sequences of moves are designed so that if you made an error (such as turning a face the wrong way, getting lost in a sequence of moves, or even dropping the cube in the midst of a sequence) you will only have to retreat one step instead of starting over. It is very frustrating to discover at the last step of a solution that the cube is as badly scrambled as it was when you started. Correcting an error is made easy because so little of the previously completed parts of the cube are disturbed. This book provides sequences of moves that are designed to deal with the most common errors.

It features easy-to-memorize sequences of moves.

It is possible to solve any starting arrangement of the cube by a few similar sequences of moves. It is easy to get used to these moves. They consist of similar sets of 3 to 5 face turns that are done repeatedly.

The required moves are set in **bold face** and are designated with a double asterisk (**) throughout. This makes memorizing the solution simpler. (Impress your friends!)

It is independent of color.

This solution is not coded by the colors of the faces. It will therefore work no matter what the 6 colors are on your cube and how they are distributed on the 6 faces. It would be nice to present a solution that read: "turn the orange face, then turn the red face," etc. However, there are various manufacturers of these cubes, and they all choose either different colors for the 6 faces or different distributions of these colors. (For example, orange is not next to yellow on all cubes.) There are 30 different ways of distributing 6 colors on the 6 faces of a cube.

Solutions that use colors (and they do exist) can only work for a specific coloring of the cube. The notation scheme used in this book is independent of color and will work for any coloring and any manufacturer's cube.

On the other hand, this solution does not use the fewest number of moves possible. It was intentionally designed for clarity and involves the least number of uncomplicated decisions rather than the fewest number of moves. There are many stages at which reducing the number of moves is possible, and these are given as short cuts. Nevertheless, with practice, it is possible to solve random arrangements of cubes in under 3 minutes.

If you would like a few hints before trying to solve a cube problem, read only pages 21 to 23.

Now enjoy your impending victory over that infernal cube!

ABOUT THE RUBIK'S CUBE:
A brief history

This fascinating puzzle was designed by Ernö Rubik, an architect and teacher in Budapest, Hungary. It was apparently designed independently by Terutoshi Ishige, an engineer in Japan. Both applied for patents in the mid-1970s. Professor Rubik designed the cube as an aid to his students in recognizing spatial relationships in three dimensions.

If you have already experimented with solving the puzzle, you may wish you could follow the movements in three dimensions more easily. It is challenging, to say the least, to establish the relative positions of the small cubes after only 2 turns of different faces.

The cubes were first manufactured in Hungary and became available in Europe in 1978. It is only recently that they have now achieved their substantial worldwide popularity (many millions sold!). They have been widely available in the United States only since 1980. It seems that these cubes were almost as common as ornaments during the 1980 holiday season.

How does it work?

The ingenious mechanism that allows all 6 faces of the cube to rotate is so simple one is tempted to ask, "Why didn't I think of it?" The difficulty is, of course, in figuring it out for the first time.

It appears that all the small cubes can move about; in fact, however, only the cubes on the corners and edges actually move. The center cubes are fixed and can only rotate in place. This is the key to understanding the mechanism.

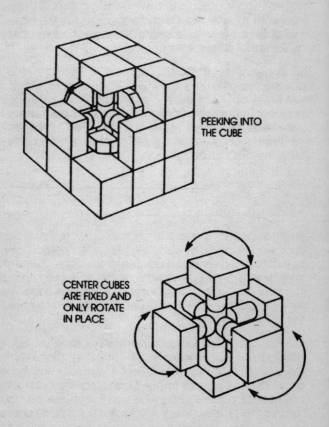

PEEKING INTO THE CUBE

CENTER CUBES ARE FIXED AND ONLY ROTATE IN PLACE

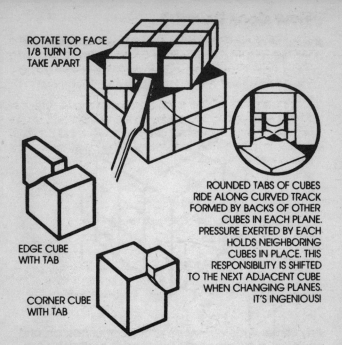

ROTATE TOP FACE
1/8 TURN TO
TAKE APART

EDGE CUBE
WITH TAB

CORNER CUBE
WITH TAB

ROUNDED TABS OF CUBES
RIDE ALONG CURVED TRACK
FORMED BY BACKS OF OTHER
CUBES IN EACH PLANE.
PRESSURE EXERTED BY EACH
HOLDS NEIGHBORING
CUBES IN PLACE. THIS
RESPONSIBILITY IS SHIFTED
TO THE NEXT ADJACENT CUBE
WHEN CHANGING PLANES.
IT'S INGENIOUS!

Each center cube is on the end of an axle. The corner and edge cubes are not attached to anything and move about the center cubes.

You may ask, "Why doesn't it just fall apart?" In fact, it can be taken apart quite easily. Simply turn one face one-eighth turn and either pull out one of the edge cubes on that face or pry it out with a screwdriver. It helps to pull the face you turned away from the rest of the cube. The remaining cubes can now be easily removed to reveal the central mechanism.

The cube does not fall apart by itself because the edge and corner cubes hold each other in place in a remarkable feat of cooperation! If you take the cube apart this way, be sure to put it back together with solid colors on all 6 faces. Otherwise, you may never be able to solve the puzzle unless you take the cube apart again.

13

Cube basics

The cube consists of 6 fixed cubes, one in the center of each face, 8 movable cubes on the corners (corner cubes), and 12 movable cubes on each edge (edge cubes). The corner and edge cubes are moved around by rotating the 6 faces.

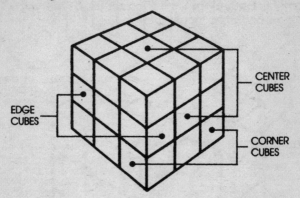

CENTER
CUBES

EDGE
CUBES

CORNER
CUBES

Any corner cube can move to any corner position, and any edge cube can move to any edge position by a sequence of face rotations.

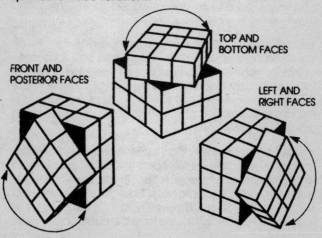

TOP AND
BOTTOM FACES

FRONT AND
POSTERIOR FACES

LEFT AND
RIGHT FACES

However, each movable cube belongs in only one location. For example, the yellow/white edge cube (my cube has one with these colors) belongs on the edge between the yellow and white faces.

The color of a face is defined by the color of the fixed center cube on that face. The yellow/orange/white corner cube belongs on the corner joining the yellow, orange, and white faces.

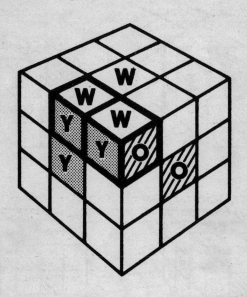

A cube is said to be placed or positioned correctly when it is on the corner or edge where it belongs. It may, however, be placed correctly and not be oriented correctly. A cube is said to be oriented correctly when it is positioned correctly and the colors match those of the adjacent center cubes. This is the desired final situation. When all cubes are positioned and oriented correctly, the puzzle is solved.

EDGES: CORRECTLY POSITIONED BUT INCORRECTLY-ORIENTED

CORRECTLY POSITIONED AND ORIENTED

CORNERS: CORRECTLY POSITIONED BUT INCORRECTLY ORIENTED

CORRECTLY POSITIONED AND ORIENTED

For example, the cubes on the left show correctly positioned edge and corner cubes that are incorrectly oriented. The cubes on the right show these edge and corner cubes correctly positioned and correctly oriented.

The method described in this book separates the problem of placing and orienting the 20 movable cubes into 5 steps. Each of the 5 steps involves the placement and orientation of 4 cubes. At any of the 5 steps, cubes that have been placed and oriented in previous steps are only temporarily disturbed.

There is no concern about cubes in later steps until that step is reached. Moreover, each cube in the first 3 sets is placed and oriented individually. In reality, then, the problem is broken up into 12 much smaller problems of placing and orienting individual cubes. It is only for the last 2 sets of 4 cubes that more than 1 cube must be placed and oriented simultaneously.

SUMMARY OF SOLUTION

A summary of the general strategy of the 5 steps follows:

1. Choose your favorite color of the 6. Place and orient the 4 edge cubes on the face with your favorite at the center.

2. Place and orient the 4 corner cubes on the top face.

3. Place and orient the 4 side-edge cubes of the middle layer. All 12 cubes placed in steps 1-3 are placed and oriented individually. The top two-thirds of the cube is now complete.

4. Place and orient the 4 corner cubes on the bottom face.

5. Place and orient the 4 edge cubes on the bottom face.

In each of the 5 steps, there are generally these 2 substeps:

1. Place the cube(s) in the proper position.

2. Orient the cube(s) correctly. This requires a temporary removal of any cube from its correct position and its subsequent return to the proper position in the proper orientation.

1. TOP EDGES

2. TOP CORNERS

3. VERTICAL EDGES
(one more in back)

4. BOTTOM CORNERS

5. BOTTOM EDGES

SUMMARY OF THE PARTITIONING OF THE 20 MOVABLE CUBES FOR
THE 5 STEPS OF THE SOLUTION. AT EACH OF THE 5 STEPS
THE 4 INDICATED CUBES ARE POSITIONED AND ORIENTED.

Why does this solution work?

The difficulty of the puzzle is perhaps made most apparent by noting that there are slightly more than 43 quintillion (4.3×10^{19}) arrangements possible for all 20 movable cubes.

How big is 43 quintillion? Well, the national debt in pennies is "only" 100 trillion. The age of the universe in seconds is thought to be only about 5 quintillion.

A very fast computer would take centuries to examine each arrangement. Yet by learning this solution and with practice, you will be able to solve the cube in a few minutes! How is this possible?

The advantage of this solution method is that by design at any time during each of the 5 steps, one is concerned with very few of the cubes involved at that step. This simplification is possible because there are fewer arrangements possible for 4 cubes (the most that are ever being dealt with at one time) than for 20 cubes. In fact, much of the time (steps 1-3), only a single cube is being dealt with. It is not difficult to keep track of just one of the 20 cubes.

In addition, the problems of positioning and orienting cubes are done separately, which significantly reduces the number of possibilities. Since there are so few arrangements of the cubes of concern at any time, it is possible to present them all and provide the appropriate action for each.

These claims can be mathematically proved, so this solution is guaranteed to be successful for solving the cube from any starting arrangement no matter how badly scrambled.

Please do not remove the colored patches from the cubes and replace them differently or take the cube apart and put it back together incorrectly! This can be done! If you do this, you will not necessarily be able to restore the cube to solid colors on all 6 faces even if you outlive the universe.

Short cuts

After you master the cube with this solution, you will probably think it is taking too long—all of 5 minutes with practice. For those on the fast track, short cuts are provided. These vary from simple sequences of moves to expert short cuts that change the overall strategy. My preferred method of solution uses these short cuts, and I can often solve random cube problems in under 1 minute. Ambitious readers may wish to improve on this further.

> Short cuts are placed in boxes at the end of each section.

Hints

● Remember that the center cubes on each of the 6 faces do not move. They can only rotate in place. This means the proper color for each face is determined by the color of the center cube on that face. For example, the face with an orange center cube must eventually be entirely orange.

● The overall strategy of this solution is given in the summary on page 18 and 19. This puzzle is too difficult to solve without some overall strategy. Try to follow the 5 steps and remember that it often helps to correctly position a cube first and then orient it so that the colors are correct.

● As you proceed in solving the cube, it is absolutely necessary to disturb temporarily some of the cubes that are placed and oriented in previous steps. For example, after one face is completed, there is no way to accomplish anything more without temporarily disturbing some of the cubes on that now completed face. You must be daring enough to be willing to temporarily undo hard-won progress in order to continue with the solution. I found this to be the most difficult part of solving the cube for the first time. For this reason, the solution in this book was intentionally designed so that very little temporary disruption of previous progress is necessary.

- Do not concern yourself with cubes that must be placed at later steps while working on earlier steps. For example, while completing the top face, don't worry at all about what happens to cubes on the bottom face. No matter how badly the cubes involved in later steps are scrambled, it will always be possible to place and orient them properly when their turn comes. There are no "dead ends."

- Simple 3-turn sequences of the type shown below are basic to steps 2–4 of the solution. You simply rotate one of the side faces one quarter turn in either direction, then rotate the bottom face, and then reverse your first rotation. The sequences of moves required to accomplish steps 2–4 are just combinations of 3 move sequences of this type. The sequence shown below changes one top corner and one vertical-edge cube without disturbing the other top-edge, top-corner, and vertical-edge cubes. A great deal can be accomplished by just performing various combinations of these sequences.

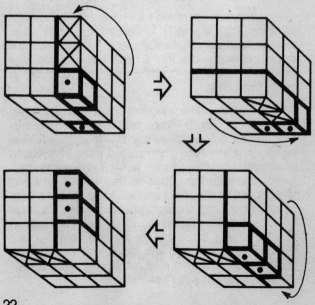

- The simple 4-move sequence shown below allows you to manipulate edge cubes without disturbing any of the corner cubes. Combinations of sequences of this type are most useful in step 5 but can be used in earlier steps (1 and 3).

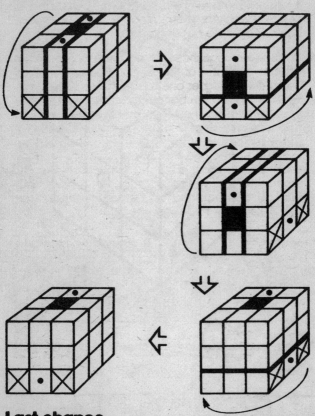

Last chance

If you wish, you may try to solve the puzzle with only the foregoing information. The remaining pages describe a complete and certain solution for any starting arrangement of the cube. Reading them might destroy the enjoyment of working out a solution yourself with just the few hints already provided.

NOTATION AND TERMINOLOGY

A standard labeling of the 6 faces of the cube is used throughout:

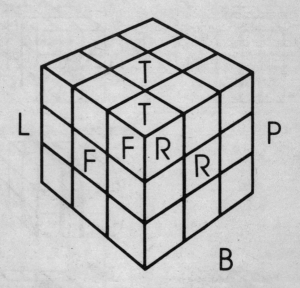

T face (top face; always the chosen favorite color)

F face (front face)

L face (left face)

R face (right face)

B face (bottom face; always one color)

P face (posterior face; rarely used)

The color of a face is the color of the center, nonmovable cube. You will choose the color of the T face and stick with it throughout. Note that the colors of the R, L, P, and F faces can vary. Thus, F can be any of four colors by simply rotating the entire cube in your hands. This then determines the colors of the R, P, and L faces.

The colors of the T and B faces remain the same after the favorite color is chosen throughout all 5 steps of the solution. The colors of the R, L, P, and F faces always remain the same during any sequence of moves but will generally differ for the next sequence performed.

THREE OF THE 6 CENTER NON-MOVABLE CUBES—CHOOSE YOUR TOP FACE AND STICK WITH IT

R+	Turn R face one quarter turn clockwise
R–	Turn R face one quarter turn counterclockwise
R2	Turn R face one half turn (either way, it does not matter)
F+	Turn F face one quarter turn clockwise
F–	Turn F face one quarter turn counterclockwise
F2	Turn F face one half turn
L+	Turn L face one quarter turn clockwise
L–	Turn L face one quarter turn counterclockwise
L2	Turn L face one half turn
B+	Turn B face one quarter turn clockwise
B–	Turn B face one quarter turn counterclockwise
B2	Turn B face one half turn
T+	Turn T face one quarter turn clockwise
T–	Turn T face one quarter turn counterclockwise
T2	Turn T face one half turn
P+	Turn P face one quarter turn clockwise
P–	Turn P face one quarter turn counterclockwise
P2	Turn P face one half turn

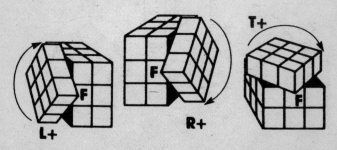

FR is the edge cube on the edge between the F and R faces at a particular time. FRT is the corner cube on the corner common to the F, R, and T faces at a particular time. The 12 edge cubes are therefore BF, BL, BP, BR, FL, FR, FT, LP, LT, PR, PT, and RT. The 8 corner cubes are BFL, BFR, BLP, BPR, FLT, FRT, LPT, and PRT. Moves and the cubes involved are described using this terminology.

In order to use the moves as written, it is necessary to hold the cube so that the cubes that are to be moved correspond to those given in the description of the move. For example, if you wish to place an edge cube in the PT position and the sequence of moves given is written for the FT position, you must rotate the entire cube in your hand so that the (former) PT position becomes the FT position. Note: by doing this, the color of the F face changes, but the color of the T face remains the same.

KEEPING THE FRONT FACE ORIENTED TOWARD YOU—ALL CLOCKWISE ROTATIONS ARE SHOWN. TO MAKE COUNTER-CLOCKWISE TURNS, SIMPLY ROTATE THE FACES IN THE OPPOSITE DIRECTIONS.

ALL ROTATIONS SHOULD BE MADE AS THOUGH YOU WERE VIEWING EACH FACE FROM THE FRONT. A CLOCKWISE TURN WOULD ALWAYS BE A TWIST TO THE RIGHT AS IF VIEWED FROM THE FRONT. SEE THE ILLUSTRATIONS BELOW.

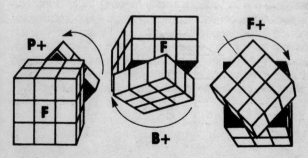

To familiarize yourself with the notation and terminology for the cubes and moves, look at the sequence of moves on the opposite page. You may wish to duplicate this sequence with your cube. I suggest choosing a dark color (e.g., blue or green) to correspond to the dark color on the drawing of the highlighted individual corner cube.

The first move is a clockwise quarter turn of the R face (cube 1 to cube 2 on the opposite page). It is symbolized by R+ and takes the highlighted corner cube from the FRT (front, right, top) position to the PRT (posterior, right, top) position.

The second move (cube 2 to cube 3) is a half turn of the T face and takes the highlighted cube from PRT to FLT.

The third move is a clockwise quarter turn of the left face. The corner cube moves from FLT to BFL.

The fourth move is a clockwise quarter turn of the B face. The corner cube moves from BFL to BFR.

The fifth move is a counterclockwise quarter turn of the F face. The corner cube moves from BFR back to FRT where it started, but note that the orientation is different.

I suggest trying them to get used to the clockwise and counterclockwise rotations for the various faces, particularly the B, R, and L faces, which are used repeatedly. No rotation of the P face is shown since it is never used. (I find it too difficult to turn this face while holding the cube nearly fixed during a sequence of moves and have therefore not included it in any moves.) Try to duplicate this sequence on your cube. Although this particular sequence is never used in the solution, it indicates how it is possible to "rotate" a corner cube to a different orientation by a sequence of moves. A trick to developing a solution is to discover sequences that do this without scrambling previously placed and oriented cubes.

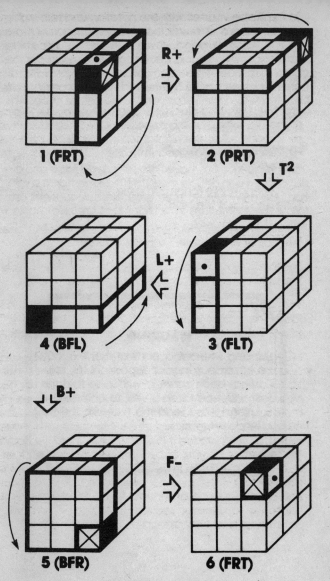

SAMPLE SEQUENCE OF MOVES

A cube is placed or positioned correctly when it is on the edge or corner with colors the same as those of the cube. A cube is oriented correctly when the colors match those of adjacent center cubes. For example, the corner cube colored orange, white, and yellow (in my case) is placed correctly when it is on the corner adjacent to the orange, yellow, and white center cubes. This cube is oriented correctly when the orange face is adjacent to the orange center cube.

To memorize a solution

Some sequences of moves are set in **bold face** and are designated with double asterisks (**). To memorize a solution, you only have to memorize these sequences and follow the instructions in sections that are also marked with a double asterisk.

WARNING

Skipping sections or parts of the text may be hazardous to your disposition! This solution has been carefully written so that the text with each step is vital to achieving a solution. Please read the text with the sequences of moves.

COMPLETE SOLUTION
STEP 1: TOP EDGES (FT, LT, PT, RT)

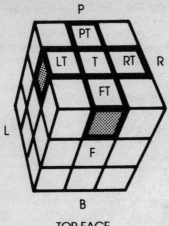

TOP FACE

To start the solution, choose a color for the T face (preferably your favorite, as you will be looking at it for some time; I like orange). Remember, the color of a face is determined by the color of the center cube only. Hold the cube so that this face is up. It remains the T face throughout the entire solution.

The objective of the first step is to position and orient the 4 cubes that belong in the 4 top-edge positions. Each of these 4 top-edge cubes is placed and oriented individually. For each of the 4 top-edge cubes, do the following 5 steps (1A-1E). If you are lucky, one or more of these top-edge cubes will be correctly positioned and oriented when you start so that you will have to do these steps (1A-1E) fewer than 4 times. If this much is not clear, reread the chapter on "Notation and Terminology."

31

1A. Hold the cube so that the FT position does not have a properly positioned and oriented cube. You may have to rotate the entire cube in your hands to accomplish this. Doing so will change the color of the F face.

1B. Locate the edge cube that belongs in this FT position. This is the desired cube.

1C. If this desired cube is currently in the FT position but incorrectly oriented, go to step 1E.

1D. Depending on the location of the desired cube, do <u>one</u> of the following 11 sequences of moves. The desired cube can be in any of 11 positions. Thus, for example, if the desired cube is currently in position RT, do the sequence RT to FT.

- Move RT to FT: R– F–
- Move PT to FT: T+ R– T– F–
- Move LT to FT: L+ F+
- Move FR to FT: F–
- Move PR to FT: R2 F– R2
- Move LP to FT: L2 F+ L2
- Move FL to FT: F+
- Move BF to FT: F2
- Move BR to FT: B– F2
- Move BP to FT: B2 F2
- Move BL to FT: B+ F2

1E. If the FT cube that is now correctly positioned is incorrectly oriented, do the following reorientation sequence:

**Orient FT: F- T+ L- T-

Remember that each of these 4 top-edge cubes is placed and oriented individually, so you may have to go through the 5 steps (1A-1E) as many as 4 times. Upon completion of this step, the T face will show a cross of the correct color of the T face (the Red Cross symbol if you chose red as your top color).

A simpler alternative strategy will work to place and orient the 4 top-edge cubes individually. This strategy is an alternative to steps 1A-1E, which you may find easier. You do not need to use this strategy if you follow steps 1A-1E.

**Alternative Method for Step 1

1. Find the desired cube (the one that belongs in one of the top-edge positions), that is not currently correctly positioned. Rotate a face (F, R, L or P) to move the desired cube to the B face. Choose whether to rotate this face a quarter turn clockwise, a quarter turn counterclockwise, or a half turn, depending on where the cube that belongs on the top edge starts out.

2. Rotate the B face until the desired cube is under the desired position.

3. Reverse the rotation done in step 1 to maintain any previously set cubes.

4. Rotate the face (F, R, L, or P) with the desired cube one half turn.

5. If necessary, reorient the newly placed top-edge cube. Use the "Orient FT" sequence given above (marked with the double asterisk).

This step is the easiest of the 5, and you may prefer to just do it your own way.

Short cut

Try to accomplish more than one goal with each move. (Think ahead!) With some careful thought and practice, you should be able to complete step 1 in less than 10 total turns of the various faces. It helps to use the T face in doing this.

STEP 2: TOP CORNERS (FLT, FRT, LPT, PRT)

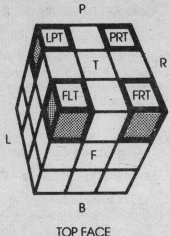

TOP FACE

The purpose here is to position and orient the 4 cubes that belong on the corners of the T face and maintain the previously placed top-edge cubes. Top-edge cubes will be moved temporarily during moves but will be properly returned.

For each of the 4 cubes that belong in one of the top-corner positions, do the following 6 steps (2A-2F). Again, if you are lucky, some of the top-corner cubes will be correctly positioned and oriented when you start, and you will have to do these steps fewer than 4 times.

2A. Locate a top-corner cube (that is, one corner cube that belongs in a top-corner position) that is not currently correctly positioned and oriented. This is the desired cube. If the desired cube is correctly positioned but incorrectly oriented, go to step 2E.

2B. If the desired cube is currently on the T face, do the following sequence. Hold the cube so the desired cube is in the FRT position.

"Move FRT to BFL: R− B− R+

This moves the desired cube to the B face.

2C. Move the desired cube, which is now on the B face, under the desired top-corner position by rotating the B face. Hold the cube so the desired position is FRT and the desired cube is in the BFR position.

2D. Do the following sequence of moves to correctly position the desired top-corner cube.

"Move BFR to FRT: R− B− R+

2E. If the FRT cube is incorrectly oriented, do <u>one</u> of the following alternative sequences.

"Orient FRT: R− B2 R+ F+ B2 F−
Orient FRT: F+ B2 F− R− B2 R+

Note that the first sequence turns the corner cube counterclockwise. The second sequence turns the corner cube clockwise. As you gain experience, you will be able to choose the proper one. However, it does not matter which you choose for now.

2F. If the FRT cube is still incorrectly oriented, repeat the <u>same</u> sequence you chose in step 2E. This will leave the FRT cube correctly positioned and oriented.

Remember that you may have to repeat these 6 steps (2A-2F) as many as 4 times to position and orient each of the 4 top-corner cubes individually. Upon completion of this step, the top one-third of the entire cube will be correct. I hope you chose a color you like for the top face; otherwise, you might be very tired of looking at it. In any case, I hope you are not seeing "red" at this stage; there is still quite a bit more to do.

Short cuts

1. There are two additional alternative sequences for moving a cube from BFR to FRT. These put the cube in an orientation different from the sequence in step 2D.

Move BFR to FRT: F+ B+ F−
Move BFR to FRT: R− B+ R+ F+ B2 F−

The short cut is to decide when to use each of the three possible BFR-to-FRT sequences. You will be able to avoid the reorientation sequence this way.

2. Move any top-corner cubes that start out on the B face up to their correct position and orientation on the T face before bringing any cubes down from the T face. Moving a cube from the B face to the T face automatically brings a different corner cube back down; on occasion, this will be one you want to bring down, anyhow.

STEP 3: VERTICAL EDGES (FL, FR, LP, PR)

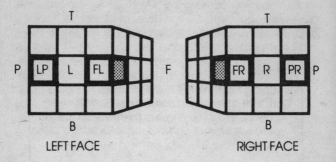

LEFT FACE   RIGHT FACE

The purpose here is to place and orient the 4 cubes that belong on the edges adjacent to but below the T face. This can be thought of as completing the "middle layer" or "equator." After completion of this step, the top-two-thirds of the cube is correct. For each of 4 cubes that belong in the vertical-edge positions, do the following 4 steps (3A-3D). Again, you may be lucky to find some of these vertical-edge cubes already positioned and oriented correctly.

3A. Locate a vertical-edge cube (one that belongs in a vertical-edge position) that is not currently correctly positioned and oriented. This is the desired cube. If the desired cube is correctly positioned but incorrectly oriented, go to step 3D.

3B. If the desired cube is not on the B face, do the following sequence of moves. Hold the cube so the desired cube is at FR.

> ****Drop FR to B face (BP):**
> **R- B+ R+ B+ F+ B- F-**

3C. At this point, the desired cube is on the B face. Rotate the B face until the color of the vertical face of the desired cube matches the color of the center cube on one of the 4 side faces. Hold the cube so the desired cube belongs in the FR position. If it matches on the R face, do the sequence BR to FR. If it matches on the F face, do the sequence BF to FR.

TURN PAGE FOR STARTING
POSITIONS FOR MOVES IN
STEP 3 (VERTICAL EDGES).

The starting positions for these sequences are shown below.

If colors match on the R face, use the sequence of moves BR to FR

"Move BR to FR:
B+ F+ B- F- B- R- B+ R+

If colors match on the F face, use the sequence of moves BF to FR

"Move BF to FR:
B- R- B+ R+ B+ F+ B- F-

STARTING POSITIONS FOR MOVES IN STEP 3
(VERTICAL EDGES)

3D. Hold the cube so the desired cube is in the FR position. If it is incorrectly oriented, do the following reorient sequence.

**Orient FR (all 15 moves):
R- B+ R+ B+ F+ B- F-
B+ R- B+ R+ B+ F+ B- F-

If the 4 vertical-edge cubes are not all correctly positioned and oriented, go back to step 3A.

Error correction

These sequences are longer than those used in the previous 2 steps. Throughout these sequences of moves, only a single top-corner cube (the one that starts at FRT) is ever more than one turn away from its correct position and orientation. Should you get lost or make a mistake during one of these sequences, simply stop and reconstruct the T face. Generally, you will have to rotate either the F or R face to return cubes to the top and then repeat one of the sequences in step 2 to return the wandering top-corner cube. Having done this, try again starting at step 3A.

Short cuts

> **1.** It is rarely necessary to drop FR to the B face (step 3B). Simply place and orient those cubes already on the B face into their correct positions. This will automatically drop another edge cube to the B face.

> **2.** (Expert) A major reduction in the number of moves required can be achieved by combining step 2 (top corners) and step 3 (vertical edges). You may have noticed that the sequences in step 3 unravel, and then partially repeat, those of step 2. Try to figure out how to combine the 2 steps.

STEP 4: BOTTOM CORNERS
(BFL, BFR, BLP, BPR)

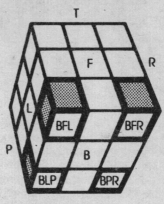

BOTTOM FACE

The next step is to place and orient the 4 corner cubes on the B face. This is done by first placing the cubes and then orienting them. These 4 cubes are not placed independently but as a set all at once. To complete this step, you must follow the instructions of steps 4A-4E just once.

4A. First, it is necessary to rotate the B face until the largest number of bottom-corner cubes are properly positioned with no concern for orientation. Ignore the edge cubes on the B face for now. Rotate the B face until you get either 2 or all 4 of the bottom-corner cubes correctly positioned. It will always be possible to position correctly either 2 or all 4 of the bottom-corner cubes by just rotating the B face. If all 4 are correctly positioned, proceed to step 4D. If 2 are correctly positioned, the 2 that remain out of place will be either on adjacent corners or opposite (diagonal) corners. If they are on adjacent corners, do step 4B. If they are on opposite (diagonal) corners, do step 4C.

ADJACENT
EXCHANGE

DIAGONAL
EXCHANGE

4B. If the 2 out-of-place corner cubes are adjacent, do the following sequence. Hold the cube so that the cubes that must be exchanged are BFL and BFR.

¨Exchange BFL and BFR:
R- B- R+ F+ B+ F- R- B+ R+ B2

Now proceed to step 4D.

4C. If the 2 out-of-place corner cubes are diagonal, do the following sequence. Hold the cube so that the cubes that must be exchanged are BFL and BPR.

¨Exchange BFL and BPR:
R- B- R+ F+ B2 F- R- B+ R+ B+

4D. The 4 bottom-corner cubes are now all properly positioned. Look at the B face and hold the cube so that the color pattern on the bottom corner cubes matches one of the 7 patterns BC1-BC7. Only the bottom color is indicated in black on the patterns BC1-BC7. The only other possibility is that all are oriented correctly. Remember to ignore the bottom-edge cubes for now. If necessary, turn the entire cube in your hand until the pattern matches one of the 7.

This matching of color patterns is done to determine which will be your F face for the next sequence. Make sure the T face is as always and proceed to step 4E.

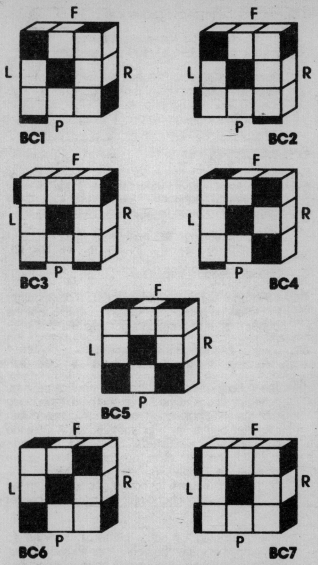

BC1

BC2

BC3

BC4

BC5

BC6

BC7

THE SEVEN POSSIBLE COLOR PATTERNS ON THE BOTTOM FACE.
REMEMBER TO IGNORE THE COLORS OF THE BOTTOM EDGE CUBES.

4E. Perform the following sequence.

"Orient bottom-corner cubes:
R− B− R+ B− R− B2 R+ B2

If the 4 bottom-corner cubes are not correctly oriented after performing this sequence, go back to step 4D. You may have to perform the sequence in step 4E as many as a total of 3 times to orient the bottom-corner cubes.

Error correction

Before doing the orientation sequence in step 4E, it is necessary to hold the cube correctly so that the T face is as always and the F face is as indicated in the drawings of the 7 patterns BC1-BC7. Should you hold the cube incorrectly and do this sequence, you will merely get back a different pattern of the 7 possible. Should this happen, simply match the pattern you do get to one of the 7 and try again by repeating step 4D.

If you make a mistake when positioning or orienting the bottom-corner cubes, it is possible you may end up with one of the vertical-edge cubes misplaced. Should this happen, you will have to go back to step 3, correct the error using one of the sequences in that step, and then return to step 4.

Short cuts

1. Use the reverse of the orientation sequence.

Orient bottom-corner cubes (reverse):
B2 R– B2 R+ B+ R– B+ R+

whenever the BC2 pattern occurs. Use the forward sequence given above whenever the BC1 pattern occurs. This will usually eliminate 8 moves from the overall solution.

2. This shorter sequence exchanges two diagonal bottom-corner cubes as required in step 4C. It does not matter which face is the F face. Only the T and B faces must remain the same as always. Use this as a replacement for the sequence in step 4D.

F– B– R– B+ R+ F+

STEP 5: BOTTOM EDGES (BF, BL, BP, BR)

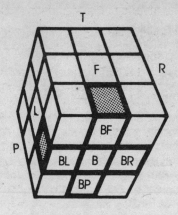

BOTTOM FACE

The final step is to place and orient the 4 bottom-edge cubes. First check to see how many of these 4 cubes are already correctly positioned (without regard to orientation). At this point, there will be either 4, 1, or none of these bottom-edge cubes correctly positioned.

If none are correctly positioned, go to step 5A.

If 1 is correctly positioned, go to step 5B.

If 4 are correctly positioned, go to step 5C.

The sequences (L– R+) and (L+ R–) are done repeatedly in this step. They are put into parentheses for clarity.

5A. Do the following sequence (all 11 moves). It does not matter how the cube is held as long as the T and B faces remain as before.

    **(L– R+) F+ (L+ R–) B2
    (L– R+) F+ (L+ R–)

47

Now there will be exactly one of the bottom-edge cubes properly positioned, although the orientation may be incorrect. At this point, go to step 5B.

5B. Hold the cube so that the properly positioned bottom-edge cube is in position BF. Then do the following sequence (all 11 moves).

$$^{**}(L-\ R+)\ F+\ (L+\ R-)\ B2$$
$$(L-\ R+)\ F+\ (L+\ R-)$$

This sequence is the same as the one in step 5A. If the 4 bottom-edge cubes are still not correctly positioned, do the sequence again with the properly positioned cube in position BF. At this point, all the bottom-edge cubes will be correctly positioned. If the cube is completed, go to step 5G; otherwise, go to step 5C.

5C. All 4 bottom-edge cubes are correctly positioned, and the B face will show the bottom color in one of the three patterns BE1-BE3. Hold the cube so the pattern matches. If necessary, turn the entire cube in your hand until the pattern you have matches one of the three. Be sure the T face is the same as always and the F face is as indicated in the patterns BC1-BC3.

If it is BE1, do step 5D.

If it is BE2, do step 5E.

If it is BE3, do step 5F.

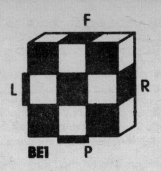

BE1

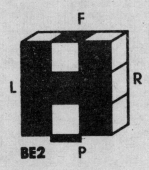

BE2

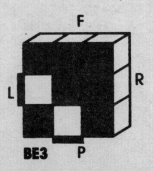

BE3

THE THREE POSSIBLE COLOR
PATTERNS FOR THE BOTTOM FACE

49

5D. Do the following sequence to complete the puzzle (all 18 moves).

(L- R+) F2 (L+ R-) B2 (L- R+) F+
(L+ R-) B2 (L- R+) F2 (L+ R-) B-

Now go to step 5G.

5E. Do the following sequence to complete the puzzle (all 30 moves). Hold the cube so that the 2 correctly oriented bottom-edge cubes are in positions BL and BR.

(L- R+) F+ (L+ R-) B+
(L- R+) F+ (L+ R-) B+
(L- R+) F2 (L+ R-) B+
(L- R+) F+ (L+ R-) B+
(L- R+) F+ (L+ R-) B2

This is a long sequence, but note that you are always turning the F and B faces in the same direction (clockwise). Go to step 5G.

5F. Do the following sequence (all 17 moves). Hold the cube so that the BF and BR are correctly oriented.

**"(L- R+) F+ (L+ R-) B- (L- R+) F-
(L+ R-) B- (L- R+) F2 (L+ R-)**

Then go back and do step 5B, after which the puzzle will be completed.

5G. Tell everyone within shouting distance and reward yourself with a trip to the refrigerator.

**Alternative method for step 5

It is possible to complete the cube from all 3 patterns (BE1, BE2, BE3) using only the sequences in steps 5F and 5B. If you get pattern BE1 or BE2, simply perform the sequences in step 5F and then 5B as directed, making sure only that the T and B faces remain the same. This will give back pattern BE3, which is then completed using step 5F.

Error correction

The net result of the sequence

$$(L- R+) \ F \ (+, \ -, \text{or } 2) \ (L+ \ R-)$$

is to move one of the top-edge cubes to the B face and replace it with one of the bottom-edge cubes. Later on, the top-edge cube is returned to its proper position and orientation. Should you make a mistake or just get lost during one of these lengthy sequences, simply return the wandering top-edge cube to its correct position by doing the following:

1. Hold the cube so that the wandering top-edge cube belongs in the FT position.

2. Move the wandering top-edge cube to position BR by rotating the B face.

3. Do the sequence $(L- R+) \ F- \ (L+ \ R-)$
This will put the top-edge cube into its correct position on the T face. However, the orientation may be wrong. Should this happen, do the following sequence:

   Orient top-edge cube:
   $(L- R+) \ F2 \ (L+ \ R-) \ B+ \ (L- R+) \ F- \ (L+ \ R-)$

All the sequences in step 5, when completed, leave all 8 corner cubes, 4 vertical-edge cubes, and 3 top-edge cubes unmolested, so a serious error is unlikely. Should something disastrous occur, simply see how much of the completed portion of the cube remains correct and go back to the appropriate previous step to correct any errors.

Short cuts

1. Use the reverse of the sequence in step 5B. Remember that bottom-edge cube BF must be correctly positioned.

$$(L- R+) \ F- \ (L+ \ R-) \ B2$$
$$(L- R+) \ F- \ (L+ \ R-)$$

Determine the changes of the bottom-edge cubes that these sequences cause. Figure out when to use the forward sequence and when to use the reverse sequence. The correct choice will usually eliminate 11 moves.

2. (Expert) It is possible to reduce significantly the number of moves required to accomplish step 5 by positioning and orienting the bottom-edge cubes at the same time. Try to design sequences of moves similar to those in steps 5A and 5F that permute the bottom-edge cubes in different ways. Then learn to recognize the arrangements of the bottom-edge cubes that require these new sequences of moves. You will likely have to recognize more than just three patterns (BC1-BC3) but will complete the puzzle in significantly fewer moves. Even at this late stage, it is still possible to place and orient one of the bottom-edge cubes independently of the other three. One approach to improving step 5 is to correctly position and orient 1 bottom-edge cube in the fastest way and then deal with the remaining 3 as a set.

AFTERWORD

I can usually solve random cube problems by using this solution with the simpler short cuts provided in about 2½ minutes. I find a considerable amount of time is spent in locating cubes and aligning the faces so that the next move can be made. The stiffness of the cube is a decisive factor in time trials.

As was pointed out in the Introduction, this solution aoes not involve the fewest number of moves possible. Nevertheless, by using all the moves provided, including the "expert" short cuts, I can average less than 100 moves for solution of a random cube problem, although occasionally more than 100 moves are required. To achieve this reduction, I have to combine steps 2 and 3, but the other steps remain the same. There is nothing sacred about breaking up the solution into these 5 steps, and a better solution may well use a different method.

An interesting pastime (or obsession) is to try to find better solutions for this fascinating puzzle. It is actually possible to rationally design sequences of moves that have desired overall effects. Try the sequence:

(L– R+) F2 (L+ R–) B2

This takes the cube that starts out in the FT position to the BP position, the cube in BP to BF, and the cube in BF to FT. It is best to try this on a solved cube to see the overall effect. Now, on a solved cube, do the following 2-move sequence.

R2 T+

This has the effect of taking a cube from BR to FT. Now, on the same cube, do the sequence:

(L− R+) F2 (L+ R−) B2

This is the same as the one given above. Then "undo" the first 2 moves:

T− R2

The overall effect is BP to BF to BR to BP. By preceding the 6-move sequence with various moves and following it with the exact reverse of those moves, you can choose to exchange any 3 edge cubes without disturbing the other 9 edge cubes and all 8 corner cubes. Using tricks like this, it is possible to move the cubes around almost at will.

A real challenge is to develop a shorter solution that is also clear, simple, and error tolerant.

Happy cubing!

OTHER GAMES TO PLAY
Speed Cubing

It is a constant challenge to try to restore the cube to solid colors in the fastest time. Keep track of your best time, and as you get better, you will find you are able to break your record often. The first time through the solution may take hours. However, you will easily improve on that each time you try. Try to beat the following times:

- 20 minutes–whiz
- 10 minutes–speed demon
- 5 minutes–expert
- 3 minutes–master of cube (M.C.)

You will probably need to memorize the solution to become an M.C. If you use the expert short cuts and memorize them, you can even do better. World class is under 1 minute. It helps to have a very flexible cube and good technique.

Competition

Try racing with someone else or in a group. To make it fair all around, occasionally exchange cubes. (Some cubes rotate more stiffly than others.) Add the scores for several runs as in a skiing competition. Naturally, the lowest total time wins.

Now, instead of starting with a scrambled cube, start with a cube with solid colors on all 6 faces. See who can create these designs the fastest. Hint: always turn opposite faces together.

Shooting Star

This can be done in 4 moves from a solid-colored cube. (answer on page 63)

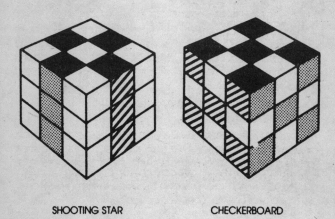

SHOOTING STAR   CHECKERBOARD

Checkerboard

This can be done in 6 moves from a solid-colored cube. (answer on page 63)

Boxes

Try to get boxes on all 6 faces. Each center cube is surrounded by cubes of another single color. This can be done in 8 moves from a solid-colored cube. (answer on page 63)

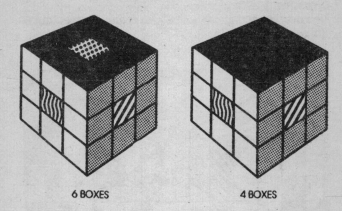

6 BOXES   4 BOXES

Now try to get boxes on only 4 faces. This is actually more difficult. Hint: if you have already gotten boxes on 6 faces, you are halfway there. This can be done in 16 moves from a solid-colored cube. (answer on page 63)

Now try these. You will not always be turning opposite faces together for these.

2-H

This can be done in 6 moves from a solid-colored cube. The H's appear on the front and posterior faces. (answer on page 63)

4-H

This can be done in 12 moves. Hint: if you have done 2-H, you are halfway there. (answer on page 63)

6-H

This can be done in 18 moves. Hint: if you have done 4-H, you are two-thirds of the way there. (answer on page 63)

From A to Z

Here is a unique challenge. Using the alphabet below, see if you can spell your name, your state, or other short words. You will find this to be very challenging, and it will test all the skills you have learned to this point and then some. Not all words are possible no matter what you do. However, most 3-letter words and many 4-letter words can be done.

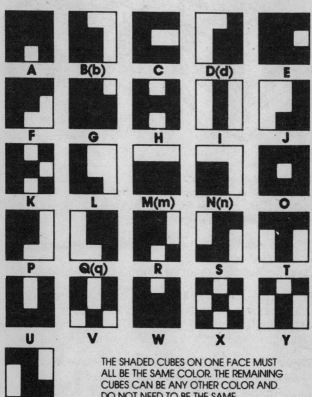

THE SHADED CUBES ON ONE FACE MUST ALL BE THE SAME COLOR. THE REMAINING CUBES CAN BE ANY OTHER COLOR AND DO NOT NEED TO BE THE SAME.

Ohio

Let's try an easy one first. Starting from a solved cube, see if you can spell Ohio as shown. Hint: use the solution for the 4-boxes pattern.

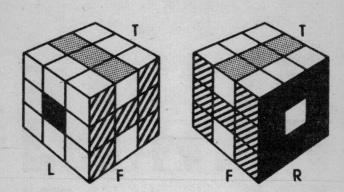

OHIO

Solution for Ohio

(L− R+) (T+ B−) (F− P+) (L− R+)
Turn the cube in your hands so that F becomes L and then do:
(L− R+) (T+ B−) (F− P+) (L− R+) R2 L2

Jack

Suppose you would like to spell your name or send a message. Many, although not all, short words can be spelled on to the cube using the alphabet, on page 59. Jack can be spelled on to the cube as shown.

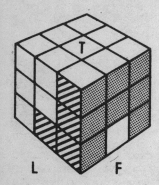 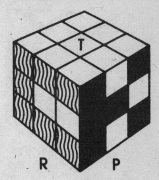

JACK

It is a major challenge to make Jack appear on your cube. The best way to do this is to use steps 1-5 of the solution (pages 31 to 53) but to place the cubes you want into the various positions rather than the one that matches the center color of the adjacent faces. This challenge is as difficult as that of solving the cube originally. However, you now have mastered the original solution and can use it to solve Jack and many other words.

Jack's solution

1. Hold your cube and look at the colors of the center cube on each face.

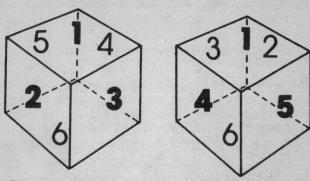

2. Write the first letter of the color that corresponds to each number on the cube shown in the box next to the number. If you are holding your cube so that red is on top, put R in the box next to 1.

1=☐ 2=☐ 3=☐ 4=☐ 5=☐ 6=☐

3. Now write the letters that correspond to the numbers in the spaces provided in the table below.

Top Edges	Top Corners	Vertical Edges	Bottom Corners	Bottom Edges
FT 26	FLT 236	FL 23	BFL 123	BF 14
LT 63	FRT 256	FR 25	BFR 125	BL 43
PT 12	LPT 143	LP 13	BLP 364	BP 45
RT 51	PRT 456	PR 46	BPR 145	BR 65

Top Edges	Top Corners	Vertical Edges	Bottom Corners	Bottom Edges
FT __ __	FLT __ __ __	FL __ __	BFL __ __ __	BF __ __
LT __ __	FRT __ __ __	FR __ __	BFR __ __ __	BL __ __
PT __ __	LPT __ __ __	LP __ __	BLP __ __ __	BP __ __
RT __ __	PRT __ __ __	PR __ __	BPR __ __ __	BR __ __

For example, if 2 is green and 6 is red, write GR in the space next to FT under "Top Edges."

FT <u>G R</u>

The colors of the center cubes are indicated in the illustration on page 62 Thus the center cube on the T face is color 1, and the center cube on the F face is color 2, and so on.

4. Now, using the colors, follow the solution from step 1 through step 5 on pages 31 to 53. Instead of putting the cube that matches the F and T faces into the FT position, put the colored cube you wrote in the table on page 62 into the FT position.

5. Do this for all 5 steps of the solution and Jack will appear before your eyes. It is particularly challenging to do steps 4 and 5 of the solution in this way. You may prefer to use the methods described in the Afterword to move some of these cubes around in specific ways.

If you can learn how to spell your name or state or some other short word, you will have your own unique puzzle to challenge your friends with.

Answers

SHOOTING STAR: L2 R2 F2 P2

CHECKERBOARD: L2 R2 F2 P2 T2 B2

BOXES (6): (L– R+) (T+ B–) (F– P+) (L– R+)

BOXES (4): (L– R+) (T+ B–) (F– P+) (L– R+)

Turn the cube in your hand so that the F face becomes the L face; then do:

(L– R+) (T+ B–) (F– P+) (L– R+)

2-H: L2 R2 B2 L2 R2 T2

4-H: L2 R2 B2 L2 R2 T2

Turn so the F face becomes the L face; then do:

L2 R2 B2 L2 R2 T2

6-H: Get 4-H. Then turn so that the T face becomes the F face; then do:

L2 R2 B2 L2 R2 T2

THE SIMPLE 5-STEP SOLUTION

1. TOP EDGES

2. TOP CORNERS

3. VERTICAL EDGES
(one more in back)

4. BOTTOM CORNERS

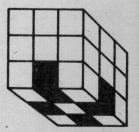

5. BOTTOM EDGES

THE SOLUTION THAT'S GUARANTEED TO WORK ON YOUR
CUBE REGARDLESS OF MANUFACTURER OR COLORS.
AT EACH OF THE 5 STEPS THE 4 INDICATED CUBES ARE
POSITIONED AND ORIENTED.